THE CALIFORNIA GOLD RUSH

A HISTORY JUST FOR KIDS

SAM ROGERS

KidCaps Books
ANAHEIM, CALIFORNIA

Contents

About KidCaps.. 1

Introduction 3

What led up to the Gold Rush?............................ 10

Why did the Gold Rush happen? 19

What happened during the Gold Rush?.................. 24

What was it like to be a kid during the Gold Rush? 35

How did the Gold Rush end?................................ 39

What happened after the Gold Rush? 42

Conclusion.. 49

ABOUT KIDCAPS

KidCaps is an imprint of BookCaps™ that is just for kids! Each month BookCaps will be releasing several books in this exciting imprint. Visit our website or like us on Facebook to see more! www.kidcaps.com

[1]

INTRODUCTION

The long car carried the men down into the depths of the Empire Gold Mine outside of Grass Valley, California. The year is 1850. These men are working to find and extract gold from the hard quartz rock in the Foothill Mountains of California. The tunnel that they were descending down wasn't very high, so they had to watch their heads. There they were, a line of men, one seated behind the other, moving slowly down a track.

One half mile (about 2500 feet) above them, their coworkers worked with a large motor that released the tension on a length of rope and thus lowered them further and further down into the black depths below. They still had another 2500 feet to travel before they reached the bottom.

Once they get to the bottom, the men are able to stand up completely. It is like another world down here in the mine, 5,000 feet (nearly one mile) below the surface. The mining here goes on twenty four hours a day. Even as they are arriving, another group of men is climbing into the car that they just got out of. A moment later, they see the little car behind them taken back up. In a few minutes, it will come back down with more workers, and the process will be repeated. The miners will be here for the next 10 hours, until their shift has ended. They turn on the lights on their hardhats and walk further into the mine.

Large pipes go past them. The pipes are connected to large steam engines on the surface. Running twenty four hours a day, these engines continuously pump out the water from the deepest parts of the mine shafts. If these pumps stopped working for even a few hours, the miners would most likely drown in the water that would quickly rise up. It can be scary down here, but it is no place to panic. A man has to keep a cool head about him. There are over 2.5 miles of tunnel down here, and it is easy to get lost. The workers stayed together and followed the noise of the others working and shouting ahead of them.

After a short walk, they saw their coworkers. Using large hammers and chisels, the men were putting deep, round holes into the tough rock on the walls of the tunnel. After they had made several holes, a different worker came and carefully placed a stick of dynamite in each hole. He placed a blasting cap on the end of each stick, which was already connected to the wires leading back to the detonating handle. The warning was given, and all of the men moved a safe distance away from the scene. As soon as everyone was accounted for, the blasting caps were detonated, causing a huge explosion. Up on the surface, explosions like this one would be heard far away from the mine, 24 hours a day, six days a week.

Once the smoke had cleared and all of the dust had settled, the miners would go back to the work area and begin to put the rocks into heavy metal carts. The carts, once full, would be pulled out of the way by donkeys that lived underground, seven carts at a time. Then, they would be pulled the 5,000 foot distance all the way to the surface. While the men below kept blasting the rock loose and digging the tunnel deeper and deeper, a few inches at a time, the men above would get to work crushing the chunks of quartz (called "ore") that had come up in the carts. They would pound the rocks into a fine dust using a stamp mill, which was a series of heavy mechanical hammers. Then, using a special chemical process, they would separate the gold flakes from the rock dust. For every two thousand pounds of rocks that they smashed and cleaned, they would get about one ounce of gold. Even that small amount of gold was enough to make them a nice profit.

Yes, it was the height of the gold rush in California, USA. What do you know about the 1849 gold rush? It was a very exciting time in the United States. People were looking everywhere for gold: in streams, underground, and in the mountains. They were even diverting entire rivers in order to get access to the dry riverbed below, to see if there was any gold there! It was a time when everybody who heard about it started thinking about striking it rich and becoming wealthy overnight.

The 1849 gold rush had small beginnings, as we will see. However, soon after it started, hundreds of thousands of people would come from all over the world to try their hand at gold prospecting and see what they could find. Friendships and alliances would be formed, and enemies would be made. Before the gold rush peaked in 1852, over $2 billion in gold will have been panned, mined, and sold. Some people will have died trying to get rich; others will have died trying to stop all of the new arrivals to California from destroying the area looking for gold. Entire ethnic groups will have become the targets of terrible treatment and outright persecution, while others will have become richer than their wildest dreams. The gold rush was a very unique time in American history.

In this report, we will be taking a closer look at this momentous period. We will find out why there was so much gold in California in the first place, and how it came to be discovered by and belong to Americans. We will have a look at why so many people came running to California, and what some of the immediate and long term consequences were. We will hear tales of success and of heartbreaking tragedy. We will learn what it was like to be a kid in those days, and about how the gold rush finally ended.

This handbook is also going to teach us some very important key terms that will help us to understand this period better. For example, we will learn about a philosophy called Manifest Destiny, about the different methods used to get gold out of the earth (like Placer mining and hard Rock mining) and about the real impact of the gold rush on history. You might be surprised when we discuss how important the gold rush really was to American history. It's safe to say that, if it weren't for the gold rush, America today would be very different.

Are you ready to start learning more? Then let's begin!

[2]

WHAT LED UP TO THE GOLD RUSH?

You might ask yourself: why was there so much gold in California? That's a great question. To find out the answer, we have to go back in time about 400 million years. Do you know where California was found 400 million years ago? It was at the bottom of the ocean. The land that we know as California hadn't yet been pushed up out of the water by tectonic plate movement. What was going on while the land was underwater? For thousands of years, volcanic activity went pushing out lots and lots of liquid magma (molten lava) all over the surface of the ocean floor (which would later rise to the surface and become California). The magma carried with it lots of minerals- including gold- which then covered the surface of the ground. Eventually, these minerals cooled off, hardened, and formed streams inside the solidified quartz rocks.

As time went by, the constant movement of the tectonic plates forced the land that would become California up to the surface, carrying the streams of gold in the quartz rocks with it. Primarily found in the Sierra Nevada mountain range, the gold was slowly eroded away by thousands of years of wind and rainstorms. It flowed down the mountainsides into streams and rivers, which later got smaller and smaller or dried up entirely. The gold, sometimes in chunks called "nuggets", other times in flakes or dust, would stay in the loose gravel on the sides of the streams. There it lay for thousands and thousands of years, until the prospectors of the Gold Rush found it.

It is amazing to think that such huge quantities of gold were just sitting there, so easily accessible, for such a long period of time. Why didn't the Native American population who had been living in California start gathering it? Well, simply put, Native Americans didn't value gold as much as other cultures (like Europeans). For them, feathers, animal furs, and shells were much more valuable and useful than small yellow pieces of soft rock. It wasn't until American settlers arrived in the mid-1800s that anyone started to notice to all of the gold literally just lying around. But how did it come to be that American settlers were in California in the first place to find all that gold?

California, previously called "Alta California", used to belong to Spain, and then to independent Mexico. It was first explored in depth by Spanish missionaries who moved steadily northward from Mexico, establishing missions and spreading their religion to the native peoples. Once they became independent, Mexico came to own a large part of what is now the United States. Their territory included California, Nevada, Utah, and Arizona. However, all of that changed with the Mexican-American War that broke out in 1846.

In 1836, the Republic of Texas had declared their independence from Mexico. During the years that followed, there was constant fighting about where exactly the border between the new Republic of Texas and the country of Mexico was. What's more, as Texas began to be considered for statehood in the United States (something that would eventually happen on December 19, 1845) Mexico began to threaten the United States. Their officials said that if the United States accepted Texas as a state there would be a war. How would the Americans react?

For reasons that we will see in the next section, the President at the time, James Polk, was very interested in pushing westward. For him, Texas was worth fighting for. During the two year war that followed (called the Mexican-American War) Mexico's soldiers lost battle after battle, most of which were fought on Mexican soil. Finally, on 2 February 1848, the Treaty of Guadalupe Hidalgo was signed in Mexico City, officially ending the Mexican-American War. In addition to ending the fighting, the treaty also included some very interesting details as far as the territory including California was involved: the land would be ceded to the United States.

That's how the United States got ownership of California. At first, it was mainly military personnel who lived there, as it was not officially a state or even a territory yet. It was simply a piece of acquired land that had to be organized and explored a little bit at a time. There were some Europeans and Americans already living there when the treaty was signed, however. For some time, Mexican officials had been allowing small groups of pioneers (men who went to isolated areas to settle them) to explore California and to set up small farms and forts there. By 1848, about 1,000 settlers were living in California, mainly in the San Francisco bay area. One such man was John Sutter.

John Sutter, original born Johann August Sutter in Germany, came to California in 1839. In August of that year, he began to build a fort that he named after himself. It was near the Sacramento River in Northern California, about 90 miles away from San Francisco. After establishing friendly relations with the local tribes of Native Americans (the Maidu and the Miwok tribes) Sutter decided to focus on farming, raising livestock, and on making a profit from the abundant natural resources to be found. In 1848, he formed a partnership with James W. Marshall to build a sawmill. This mill would process local trees into usable wood for later sale (and a nice profit). The sawmill, built near the modern day town of Coloma, would be powered by the American River that it was built beside.

Because of his experience, James Marshall was in charge of the actual mill construction while Sutter stayed in his fort and focused on his business there.

While looking over the progress of the mill on the morning of January 24, 1848, Marshall noticed some shiny yellow flakes trapped in the lower part of the mill where the river water was flowing through. He showed them to Sutter, and they did some basic tests on the metal to see what it was. The test confirmed it: they had found gold.

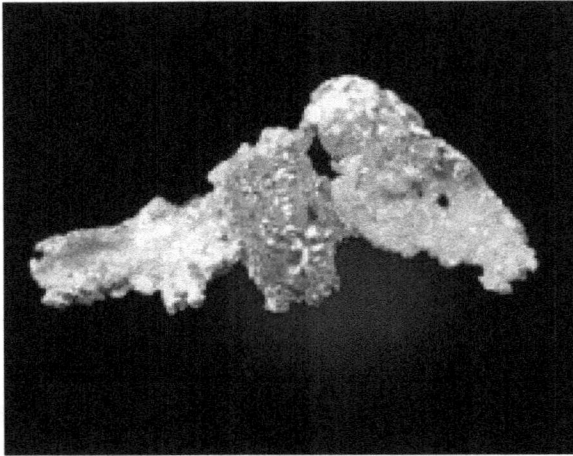

At first, no one could believe that gold was just sitting there, waiting for someone to go and pick it up. Most of the people who heard about the story (even though Sutter and Marshall tried to keep it a secret) thought it was just a hoax or a rumor. Sutter started trying to buy up the land around the mill so that he could legally claim the gold, and Marshall decided to keep working on the sawmill itself. However, news of the gold discovery could not be contained for long. Marshall's men began looking for gold during their breaks. Another man from nearby San Francisco (which was called "Yerba Buena" in those days) walked up and down the streets, telling the 1,000 or so inhabitants that gold had been found in the American River. This man, Samuel Brannan, wasn't as interested in finding gold himself as he was in selling supplies to interested miners. He became one of the richest men during the Gold Rush, most of his money coming from being a good salesman with an eye for opportunity.

As ships came into the San Francisco port, they travelled back to the east coast, and the passengers told the people there all about the massive amounts of gold in California. In the meantime, nearby residents from Oregon and Mexico began heading to California to get some gold for themselves. By August, New York newspapers were reporting Marshall's gold discovery. In December, President James Polk spoke about and confirmed the discovery to Congress during his State of the union address. Admitting that even he himself thought that the first reports were a hoax, he then said:

> "The explorations already made warrant the belief that the supply is very large and that gold is found at various

places in an extensive district of country."[1]

This official confirmation from the President of the United States was enough to make thousands upon thousands of people leave everything behind and make the trip to California. The Gold Rush had begun.

---

[1] James Polk gold quote: http://www.findingdulcinea.com/news/on-this-day/On-This-Day--President-Polk-Sparks-the-California-Gold-Rush.html

[3]

WHY DID THE GOLD RUSH HAPPEN?

As you can imagine, a find of this much gold would make everyone living nearby get very excited. Sutter himself had to have been very happy to think about the gold that was on his land. However, the question that we want is answered is this: why did people go so crazy about the discovery of gold? Nowadays, if you hear about someone finding oil, gold, or diamonds, do you pack up everything and run to try to find some for yourself? Of course not! Then what made everyone back then leave their entire lives behind to go and look for gold? There were two main reasons why the Gold Rush happened like it did: the issue of land ownership in California and the attitude of Americans at the time. Let's look closer at these two factors.

Land ownership in California. As you may re-call from the previous section, California in 1848 was not yet a state or even an organized U.S. territory. It was simply a piece of land ac-quired by the government that had yet to be explored and divided. There was no govern-ment, no courts, no police, and no land owners. When gold was discovered, there was no one to decide who could work on what piece on land. In other words: it was a free-for-all. The first person who arrived could "stake his claim" (fix his own land boundaries) anywhere he wanted. If he didn't see results, he would move on to another area and do the same.

This idea, the notion of arriving and setting up camp anywhere you wanted with gold just waiting to be found, captured the imagination of people across the world. It was just too good to pass up. However, land ownership is-sues weren't the only factors that contributed to the Gold Rush. There were others.

Attitude of Americans at the time. When the first settlers came to America and established the Jamestown colony back in 1607, they quickly learned that the best way to make money in the New World wasn't any sort of get-rich-quick scheme; the best way was spending long years doing hard work and showing determination. For centuries, that was the American dream: buy your land, build your house, raise your family, and build your fortune over a lifetime.

The Gold Rush changed all of that. The new American dream became getting rich overnight by striking it rich in California. This attitude helped to fuel the fire that became the Gold Rush. However, there was something else involved, something that involved President James Polk himself: a philosophy called "Manifest Destiny". Have you ever heard of that philosophy before?

As you know, American colonists first settled on the East coast of the United States. As time went by, they settled northwards and southwards, and then began to push westward. But there were a few obstacles to their westward advancement: a lot of the land west of the Mississippi River already had owners: nations like France, Spain, Mexico, and Britain, not to mention the hundreds of thousands of Native Americans who had lived there for generations, who would defend their land ownership. However, in the American psyche (way of thinking) there was something very special about the land of North America: it was destined to belong to the United States, coast to coast, sea to shining sea. No one else could ever be its owner.

Although the idea had been around for a while, one journalist named John L. O'Sullivan really explained this philosophy clearly in an article he wrote talking about ownership of Oregon on December 27, 1845, a few years before the Gold Rush started. He said:

> "And that claim is by the right of our manifest destiny to overspread and to possess the whole of the continent which Providence has given us for the development of the great experiment of liberty and federated self-government entrusted to us."[2]

This philosophy, which came to be called "Manifest Destiny" after the phrase coined by O'Sullivan, came to influence some of the important decisions made by President James Polk himself. During his inaugural address in 1845, he said:

> "...It is confidently believed that our system may be safely extended to the utmost bounds of our territorial limits and that as it shall be extended the

---

[2] John L. O'Sullivan quote source: http://en.wikipedia.org/wiki/Manifest_destiny#cite_note-3

bonds of our Union, so far from being weakened, will become stronger..."[3]

This attitude of Manifest Destiny shaped his presidency. How so? For example, as we saw earlier, President Polk was not afraid of going to war with Mexico over the state of Texas. Why not? Because he felt that the United States was destined to own land all the way to the Pacific Ocean, including Texas. The Treaty of Guadalupe Hidalgo that ended the Mexican American War made Polk and other Americans very happy because it was giving them the land that they felt America was already entitled to.

When gold was discovered in California, this little frontier outpost became the focus of worldwide attention. However, Americans themselves especially felt a special desire to go. After all, it was their destiny to get all the riches that the New World had to offer, including the gold of California.

As we have seen, several factors (not only the value of gold itself) contributed to making to the Gold Rush such a large phenomenon. What actually happened during the Gold Rush? Let's find out.

---

[3]   Source   for   James   Polk's   quote: http://www.learningfromlyrics.org/manifest.html

[4]

WHAT HAPPENED DURING THE GOLD RUSH?

During the Gold Rush, the population of California exploded. It went from about 1,000 settlers to over 300,000 within a few short years, with some 90,000 prospectors coming west in 1849 alone. Cities were built almost overnight, and sometimes they were abandoned just as quickly. Local governments were quickly organized and began to bring some law and order to the chaotic lifestyle of the miners and prospectors. Let's have a closer look at some different aspects of life during the Gold Rush. We will look at the following areas:

- The people who came to find gold
- The methods used to find gold
- Some problems faced by prospectors
- Surprising facts about the Gold Rush

Are you ready to learn more about what happened during the Gold Rush? Then let's continue.

The people who came to find gold. James Marshall found gold at Sutter's Mill in January of 1848. Within a short time, settlers living in San Francisco and as far away as Oregon heard about the discovery and came running, picks and hammers in hand, to claim their share. The ones who arrived first, in 1848, found ideal conditions for finding gold. The gold was easy to locate, and required very little work to get to it. There were hardly any other prospectors around (meaning there wasn't a lot of competition) and most of the gold hadn't been found yet.

For the prospectors who came in 1849 (called the forty-niners) things weren't quite as easy. By the time they arrived, most of the good pieces of land had already been claimed. The gold that was just sitting in the gravel along the rivers and streams had already been found, and there were lots more people now fighting to find their share. In fact, the large amount of people moving to California had raised the prices of everything in the cities. San Francisco had grown so much that they actually ran out of space to build any more new houses.

A solution that they came up with was to create a landfill of garbage and rocks and to dump it into the bay. This was done, and it provided more surface area for houses and shops.

Other towns sprung up almost overnight. These towns, called "boom towns" would form around an area that was thought to have a lot of gold. Prospectors would come running and merchants would come running after them. Houses and shops would be built, and the town would bustle for a few years or maybe a little more. However, after the gold supply had run out, the prospectors would move on, and the merchants would go with them. The town would suddenly become very quiet. The large empty buildings and the lonely, deserted streets became a familiar sight during the Gold

Rush, and these places came to be called "ghost towns".

The people who came to look for gold also had to face all kinds of dangers. These dangers began on the actual trip out to California. In 1849 there was no railroad service to California (and of course no airplanes) so the only way to get to the gold was either by land (with horse and wagon) or by sea. In fact, it was divided about fifty-fifty: about half came by sea and the other half by land. To come by sea meant getting on a ship on the east coast and making the dangerous trip down around the southern tip of South America. For those who weren't brave enough to try that route, they could always sail to the Isthmus of Panama, cross the overland route on donkeys (or walking) and wait for a ship headed north to the port of San Francisco.

The overland route meant crossing the Rocky Mountains and then the Sierra Nevada mountain range, which was not an easy feat even for experienced travelers. Some who had recently tried to cross the Sierra Nevada got stuck in deep snow for the entire winter of 1846 in the Donner Pass. With very little food and no hope of rescue, this party became notorious among settlers because some of them actually resorted to eating fellow dead travelers. Stories like this made some people think twice about travelling over land. No everybody was deterred, and thousands kept on going west because they wanted that California gold.

Once arriving in California, there were different types of dangers to worry about. There was disease, cave-ins at gold mines, violence in the streets, and so on. In fact, one out of every twelve people that went west during the Gold Rush died within a short time. It was not an easy way of life, but for many it was worth it to get as much gold as they could.

The methods used to find gold. At first, there wasn't a lot of technology required to find gold in California; it had more to do with being in the right place at the right time and with being willing to do lots of hard work. Let's look at some of the methods used during the Gold Rush to collect that gold.

- Placer mining. This was the first and most popular type of mining during the Gold Rush. "Placer" is from a Spanish word that refers to the "shoal", or shallow banks, of a river. It was in these gravel-covered shoals that many of the loose gold nuggets and leaves were to be found. There were several ways that a miner could look for gold in these areas working by himself: he could use a pan with ridges to swish the gravel around, letting the heavy gold settle at the bottom, or he could use a "rocker" to sift through larger quantities of gravel.

- Hydraulic mining. After all of the gold near the rivers and creeks had been found, prospectors had to start looking harder for California's gold. First used in 1853, this method proved to be one of the most productive; and the most destructive. By the time that it was banned in 1884, an estimated 11 million ounces of gold were discovered and sold. How did hydraulic mining work?

The goal of hydraulic mining was to loosen the gold that might be buried in hillsides. Large pumps were brought in to blast hillsides with strong streams of water. The dirt and rocks from the hill fell to the ground and were carried away to a sluice (a miniature aqueduct) where the heavy gold would fall to the bottom. This method was banned in 1884 because of the large quantities of dirt and ore that were washing into the San Francisco Bay.

• Hard rock mining. This is the method that we saw described in the introduction, where men would use dynamite to blast deep into the earth, looking for strains of gold hidden in hard rock (which was usually quartz in California). This was also a dangerous method for the miners. The tunnels could collapse at any moment, dangerous gases could suffocate them, or the water pumps could stop working and they could all drown in the groundwater. Despite the danger, some 60% of the gold extracted during the Gold Rush was from hard rock mining techniques.

Some of the problems faced by prospectors. As if it weren't hard enough to travel thousands of miles and to work ten hour days one mile underground, prospectors had other problems to worry about. For example, do you remember who was living in California before the thousands upon thousands of gold seekers arrived? Well, there were the "Californios", the former Mexican citizens now living in the United States territory known as California. But there were also thousands and thousands of Native Americans, organized into various tribes. In 1845, there were about 150,000 Native Americans (although there would only be 30,000 by 1870). For generations, these tribes had lived off of the land and had fished in the waters. Now, this was the same land and the same water that were being taken away from them or that was being polluted by prospectors. How do you think the Native Americans felt?

As you can imagine, the native people (both Native Americans and Californios) did not appreciate being pushed around and kicked off of their lands. Many Native Americans, away from the only lands and food sources that they had ever known, died of starvation. Others resisted and fought back against the prospectors, which brought even greater violence. For example, one settler named Colonel John Anderson was killed in the spring of 1852 in Northern California by a local Wintu Native American group. The group stole his livestock (most likely to feed themselves after having lost their hunting grounds) and ran away. Soon after the murder, 70 men led by Sherriff William H. Dixon hunted down the local Wintu tribe and killed all but three (small children) of the 150 member tribe. As they got closer, however, they saw that the murderers of the Colonel were not among the dead. They had killed the wrong people.

Fights like these (between settlers and Native Americans) were common, and it was the Native Americans who usually ended up losing. That being said, it was also common for the prospectors of different races and nationalities to fight among themselves. Coming from so many different nations, including China and Australia, non-American gold seekers were taxed heavily and taken advantage of.

Surprising facts about the Gold Rush. While newspaper reports often talked about how much gold was found during the Gold Rush, and how many prospectors struck it big, gold seekers weren't the ones who made the most money during those years- it was the merchants who really filled their pockets. As an example, do you remember the San Francisco business named Samuel Brannan who publicized Sutter and Marshall's discovery of gold? Seeing an opportunity (and predicting the coming Gold Fever) Samuel went out and bought up all of the mining equipment in the whole city. Then, he happily sold it to interested prospectors- for a much higher price than he paid for it. He became San Francisco's first millionaire.

Like Samuel Brannan, many other merchants (not only in San Francisco but also in the many other boomtowns) got rich by taking care of the needs of gold seekers. Some built hotels and lodging houses while others focused on sewing, cleaning laundry, providing entertainment, or even on making and selling clothes. For example, cotton blue jeans had become very popular during the Gold Rush because they were durable and easy to take care of. Beginning in 1873, Levi Strauss made his fortune selling high-quality work pants to forty-niners in San Francisco.

Another surprising fact has to do with the number of women and children that participated in the Gold Rush. Although we normally think of men working hard to get gold, there were also entire families that spent their days placer mining alongside streams and rivers. Later on, with the population boom, women came west to find adventure or to work. Because prospectors valued a woman's touch on their clothes, such women would sometimes set up a business and charge high prices to do a quality job of washing and ironing a man's clothes. Some of these women were widows, others had their husbands, but they all loved the life that came from living in the Wild West.

[5]

WHAT WAS IT LIKE TO BE A KID DURING THE GOLD RUSH?

As we have seen so far, living during the Gold Rush meant a lot of hard work and a lot of danger. Can you imagine seeing so much danger and death around you? Well, have you ever known someone that died? Back then, remember that one out of every twelve people who went to California during the Gold Rush died. Imagine that today: out of every twelve people that you know, one dies suddenly. It would have been terrible to see so many people getting into accidents and getting sick.

More than just worrying about other people, however, you would also have had to worry a lot about yourself. For example, what would you have done if you found a lot of gold?

Would you have been able to trust anybody to help you work the area, or would you have been too afraid to tell anyone about it? What would you have done with the gold once you had dug it up out of the ground? After all, there were lots of thieves back then. Can you understand how the Gold Rush made some people less trusting of others?

Another concern was establishing good relationships with Native Americans. It may sound easy, but there was a big problem for prospectors who were looking for gold: Native Americans were living on much of the land that had lots of gold on it. What would have been the best way to handle the situation? A lot of prospectors ended up using force to remove the Native American tribes from their lands, while others used outright violence. Do you think that it was right of them to do that? What would you have done? While it may be easy to think that we would have just made friends with the local Native American tribes, what if they didn't want to leave? At what point does a desire to be rich become more important than the way that we treat other people? Because of their dream of getting rich quickly, a lot of prospectors did some pretty terrible things to other people.

Being so far away from civilization, it was not uncommon to see people get really sick from strange diseases. When you are sick, do you go to the doctor? Of course, you do, we all do! But imagine living far away from the nearest city. If you get sick, there may not be anyone to help you. What if you came across a bear or a mountain lion in the woods? What would you do? These were all very real dangers for the people looking for gold in California.

Kids living back then were also used to work-ing hard. Think about the deep gold mine that we learned about in the introduction: Empire Gold Mine near Grass Valley, California. That is a real mine, and you can still visit it today and see what it was like to work there. Did you know that there were donkeys that spent their entire lives down inside that mine, working long hours and carrying heavy carts full of rocks and gold? Well, there were some men who spent their lives like those donkeys. They spent long days underground and got tired, dirty, and sweaty trying to get rich. Even if they found lots of gold, it was the owners of the mine who would usually keep most of it; for the workers themselves, it was just another day on the job. The American Dream of striking it rich disappeared after the first few years of the

Gold Rush; after that, looking for gold was just another type of work, like a being in a factory.

Would you have liked to be alive back then? Although there was a chance that you might get rich overnight, most people only made enough money to just get by. A lot more went home even poorer than when they started. Their dreams of striking it rich in California never came true.

[6]

How did the Gold Rush end?

Simply put, the Gold Rush ended when the gold ran out.

Throughout the years that it lasted, the prospectors had tried different methods as the gold got harder and harder to find. They started with placer mining in order to get the gold that was easy to find (buried in the gravel of river banks). Then, they began using hydraulic mining to blast away entire hillsides to get to the gold buried beneath. After that, others would go downstream to dredge through the ore and rocks that had been removed by past operations, just in case any gold had slipped through. Finally, prospectors moved on to hard rock mining. This method continued to be used to find gold in California for almost one hundred years (until the 1950s) when most of the mines were shut down because they were no longer profitable.

Because of the vast amounts of money coming from California, it was admitted as into the Union in 1850, as a free state. Billions of dollars were shipped out (in the form of gold) to the east coast, much of it ending up in the hands of the government.

Many of the people who had migrated to California to look for gold ended up staying and finding other ways of making money. Some opened schools, became politicians, started businesses, and built permanent homes. Instead of being a bunch of rowdy prospectors looking for fun and a quick dollar, Californians became like citizens of any other state: families, tradesmen, farmers, and so on.

The large migration of people to California had also done a lot to open up transportation and communication with the area. The Pony Express (mounted riders who would carry correspondence and packages all the way from Missouri to San Francisco) operated from 1860 to 1861, after which telegraph wires were installed. Soon after that, railroad tracks would follow (starting in 1863), and many of the same workers that had worked hard looking for gold would work hard blasting tunnels in the Sierra

Nevada and Rocky Mountains and laying miles upon miles of train tracks.

For example, many of the Chinese workers that had come to California looking for gold were hired to help build the transcontinental railroad. They worked long hours both in the hot sun and during the freezing winter. Some lost their lives during huge dynamite explosions in the tunnels while others were buried in avalanches in the two mountain ranges. Their labor made the railroad a success, even finishing it about seven years ahead of schedule. If it weren't for the Gold Rush, these Chinese workers would never have come to California.

The Gold Rush had ended, but its effects would last for a very long time.

[7]

WHAT HAPPENED AFTER THE GOLD RUSH?

Any time you deal with large amounts of people and money, you can be sure that important things will happen as a result. Some of the effects of the Gold Rush were felt right away, others a little later, and some are still seen today. Let's see what happened after the Gold Rush, and how it changed a lot of things about the United States.

One of the first effects was a very negative one: many Native Americans died. This happened in a variety of ways: disease, relocation, outright violence.

Disease: When settlers came to California searching for gold, they found about 150,000 Native Americans living and hunting and gathering and raising their families, just as they had been doing for countless generations. However, the first meetings between settlers and Native Americans ended up with many of the Native Americans dying of strange new diseases. How did that happen? As you may know, our bodies protect us against diseases by being exposed to just a little bit of it, especially in the cases of viruses. However, if a person is exposed to large quantities of a new virus, the body often cannot fight off the disease and the person ends up dying very quickly. This is what happened to large numbers of Native Americans in California. With all of the new settlers arriving from so many countries with so many different types of viruses, the immune systems of the native population just could not handle all of the strange new diseases, and they died in large numbers.

Relocation: As we saw earlier, Native Americans were already living on lands that were rich with the very gold that prospectors wanted. When they arrived in beginning in 1848, some of these prospectors forced whole tribes to leave these lands, the only ones that they had ever known. The Native American tribes had to go to new places where they couldn't find food, couldn't raise crops, and had no place to care for their animals. Being forced from one place to another, many Native Americans actually ended up starving to death, with no one to help them. Can you believe that some people could have been so cruel like that?

Outright violence: As tensions between the two groups got stronger and stronger with each passing year, some Native Americans began to retaliate, or fight back, against the harsh and unfair treatment that they were receiving. As you can imagine, this only made those hateful prospectors angrier and crueler. We already saw what happened at the Bridge Gulch Massacre. However, that was not the only violence against Native Americans that happened during the Gold Rush. Another massacre that took place in 1850 was called the Bloody Island Massacre. This happened when a group of enslaved Pomo people revolted and murdered their cruel masters. In retaliation, almost 25% of the group (about 100) was killed by a regiment of the United States cavalry. Sadly, there were other massacres like these.

The terrible treatment of Native Americans was one of the terrible effects of the Gold Rush. Although there had been around 150,000 Native Americans in California in 1845, before gold was discovered, there were only 30,000 by 1870, only twenty five years later.

Another long term effect was unfortunately also a negative one. Do you remember some of the methods that were used by prospectors trying to find gold? After they had found a lot of gold using placer mining, they began to use hydraulic mining methods. This is when they would blast entire hillsides with powerful streams of water to knock the rocks (and the gold inside it) loose. As the dirt, rocks, and gold fell into the streams of water below, the heavy wood would be caught in the sluices set up by the prospectors, and the sand and gravel would be washed away into the rivers.

As all of this sediment moved down the rivers, much of it ended up either in the Sacramento Valley or in the San Francisco Bay area. The changes in the river paths and the raised level of the ground had a sort of domino effect that led to major flooding each spring, which absolutely devastated towns in the Sacramento Valley. What's more, the hillsides themselves lost valuable topsoil and, as a result, nothing could grow there anymore. Sadly, these effects have lasted to this day. In the picture below, you can see a recent picture taken in Northern California of a former hydraulic mining site.

Also, some of the methods of extracting gold used dangerous elements to make it easier to separate from the surrounding water and dirt. For example, one website explains how mercury, a dangerous element in large quantities, was used in gold mining:

> "To enhance gold recovery from hydraulic mining, hundreds of pounds of liquid mercury (several 76-lb flasks) were added to riffles and troughs in a typical sluice. The high density of mercury allowed gold and gold-mercury amalgam to sink while sand and gravel passed over the mercury and through the sluice. Large volumes of turbulent water flowing through the sluice caused

many of the finer gold and mercury particles to wash through and out of the sluice before they could settle in the mercury-laden riffles. A modification known as an undercurrent (fig. 5) reduced this loss. The finer grained particles were diverted to the undercurrent, where gold was amalgamated on mercury-lined copper plates. Most of the mercury remained on the copper plates; however, some was lost to the flowing slurry and was transported to downstream environments."[4]

This mercury later went on to contaminate streams and even the ocean, where it is still present today. For this reason, we have to be careful how much fish we eat and even what kinds, because of the mercury used during the Gold Rush.

Another effect of the Gold Rush was the large amount of money received by the Federal Government. At the time, money was printed based on the amount of gold in government safes. As they received more gold, more money was printed, and spending power of the average American increased. Some of this money from the Gold rush later went on to supply

---

[4] Mercury mining source: http://pubs.usgs.gov/fs/2005/3014/

weapons, uniforms, and equipment to the Union Army during the Civil War that broke out fifteen years later. How important were these extra funds? The superior funding of the Northern Army was one of the deciding factors of the Civil War, something which completely changed U.S. history and led to the abolishment of slavery. If it had not been for the Gold Rush and all of the money that it brought to the U.S. government, who knows if the Civil War would have ended differently?

Finally, another long term effect of the Gold Rush was the diversity that we see in California today. This is perhaps one of the best results, bringing together so many different types on people together in one place. It has contributed to making California a popular place to go for art, entertainment, agriculture, and culinary artistry (which means good food for us).

These effects, whether short term or long term, were all directly related to the discovery of gold in 1848 and of the Gold Rush that followed.

[8]
CONCLUSION

When James Marshall found a few flakes of gold while working on Sutter's Mill that January morning, do think that he had any idea what would be the consequences? Could he have possibly imagined that his discovery would indirectly lead to a population boom in California, to the deaths of thousands of Native Americans, to the invention of new technologies, to the construction of the Transcontinental Railroad, or that it would even affect the outcome of a future Civil War? Is there any way that he could have known that thousands and thousands of people would have their lives turned upside down- sometimes for good, sometimes for bad- by this one little discovery? Oddly enough, James Marshall never really benefitted from his discovery, and even died a poor man some years later.

The Gold Rush brought out the best and worst in the people who participated. It brought out great business skills in men like Samuel Brannan; but it also brought out the ruthless cruelty in those who committed massacres against both Native American landowners and against Chinese miners. The desire to make money even led to unfortunate environmental consequences which we are still suffering today.

When you look at the overall picture, what do you think: was the Gold Rush a good thing or a bad thing for the United States and for the world? If you had been alive back then, would you have been one of the people running to find gold of your own or would you have been content to let others go and just to hear about their adventures?

Important things often start with little events. And as was the case with the Gold Rush, it was the first people who acted who benefitted the most. What about you? Will you be quick to take advantage of an opportunity when you see it? Will you try to see the big picture and look at the possible consequences of your actions? Will you think about other people before making big decisions? These are all val-

uable lessons to be learned from the Gold
Rush, the race that changed the United States.